The Energy of the Future

WATER MOLECULE POWER REACTOR SYSTEM

Jet Engine Application and Analysis

The Future of Planetary and Inter Planetary Travel

Cornelio Jeremy G. Ecle

Cherlowen A. Bolito

Jelinda T. Macasusi

John Mark N. Gonzales

1st Special Edition
Innovative Research Initiative and Beyond

January 2021
Systenext Alpha Research and Innovations Limited © 2021

WATER MOLECULE POWER REACTOR SYSTEM

Jet Engine Application and Analysis

Cornelio Jeremy G. Ecle

Cherlowen A. Bolito

Jelinda T. Macasusi

John Mark N. Gonzales

Copyright © 2021 Cornelio Jeremy G. Ecle

All rights reserved.

Water Molecule Power Reactor System
Jet Engine Application Analysis
By Cornelio Jeremy G. Ecle, Cherlowen A. Bolito, Jelinda T. Macasusi and John Mark N. Gonzales

All rights reserved. No part of this book may be reproduced, stored in a retrieval system or be transmitted, in any form or by any means, electronic, mechanical, photocopying, recording or otherwise, without the prior written permission from the authors or the publisher, except for in the use in academic instructional examples and research references.

Warning and Disclaimer

This book is a publication of research related works which is therefore theoretical and analytical in nature. In this view the subjects mentioned in this book may change upon the discovery of new technological breakthrough in the research and development world. The readers are advised to verify, take extra precautions and consult the experts before performing the procedures mentioned in this book.

Hydrogen and oxygen gas are highly flammable gases which must be given due respect otherwise it will cause damage to lives and properties.

The authors cannot be held responsible to any person or entity, for any damage or loss arising from the information and in reference to the technical details mentioned on this book.

Systenext Alpha Research and Innovations Limited © 2021
Copyright © 2021 All Rights Reserved.

R5.01.01.2021.SARIL.B5.2021

ISBN-13 : 979-8583994106

Dedication

The author wishes to give thanks to the Creator for making this research work possible. An enormous and significant time has been dedicated in the full completion of this research work and the efforts for this research initiative continuous through time thus, the author acknowledges the gift of time and opportunity given by the great and almighty Creator.

Table of Contents

Author's Note……………………………………..…. 1

Chapter One……………………………………….. 2

Fuel Air Ratio……………...………………….…. 6

Fuel Air Equivalence Ratio……………….……...16

Air Fuel Equivalence Ratio…………………….. 18

Review Questions…………………………..….. 20

Chapter Two………………………………….....21

Device Efficiency….......…………….......………. 22

Coefficient of Performance……………………............. 24

Review Questions……………………………..….. 25

Chapter Three…………………………………….27

Thrust Force…………………………………............28

Wind Turbine Power Analysis…………………........29

Review Questions ……………………………………. 32

Chapter Four………………………………….. 35

Resistance of a Cylindrical Material Analysis………..…...36

Volume of a Cylinder Analysis……………………........37

- v -

Volume of the Fuel Analysis... 38

Area of the Rods……………………………………….. 39

Resistance of the Rods…………………………………….. 39

Total Resistance……………………………………....40

Current Requirements…………………………….......... 43

Voltage Requirements………………………….............43

Quantity of Electrons Formula……..…...............................46

Review Questions……………………………………... 50

Chapter Five……………………………………..……52

Net Thrust Force... 54

Gross Thrust Force…………………………….....……55

Review Questions…………………………………….. 59

Chapter Six…………………………………….....…61

Input Power Requirements...63

Output Power Requirements...… 63

Review Questions……………………………………... 69

Glossary………………………………………...…. 71

Safety Precautions and Warning............................... 72

References.. 73

Author's Short Biography....................................... 74

To the Creator.

AUTHOR'S NOTE

It is the nature of man to understand and to seek what it does not understand.

Through the application of science based on facts and through the study of physics and mathematics humans have in many ways evolve. For evolution in the natural course of nature, it is within the nature of man to evolve in its ways. For man cannot rest without seeking the truth, the truth of its being.

For like the creator man had been endowed with the gift of wisdom to understand the nature of its being.

The technology that today man had develop will continue to evolve without ending for it is in the nature of man to seek what is best for itself.

In this book the study on making the Water Molecule Power Reactor System be able to utilize the technology of jet engine in order for it to be able to produce electricity is being examined in this book. This is another milestone for human kind.

CHAPTER ONE
Air to Fuel Ratio and Fuel Air Ratio

> *This chapter will examine the fundamental mathematical principles of the Air and Fuel Ratio and Fuel Air Ratio relative to its application and use on the Water Molecule Power Reactor System.*

Analysis of the air to fuel ratio and fuel to air ratio capability of the proposed Water Molecule Power Reactor System

The fundamental principles of using hydrogen and oxygen gas as a fuel to a given model of a jet engine is a very interesting subject to tackle upon and perhaps by the application of enough and sufficient scientific knowledge this technology will evolve perfectly and will find a way to human consumption.

Of course the hydrogen and oxygen gas will be produced by the process of electrolysis of sea water in the system this book calls as the Water Molecule Power Reactor System.

In the previous studies many variables had already been defined, mathematically verified and was illustrated through a writing of a book which was a precedent to this current script and research study.

In this current study this book will define and illustrate the variables needed to obtain an integrated system which will use a Jet Engine System and the Water Molecule Power Reactor System and this system integration will produce a hydrogen and oxygen gas that will power the jet engine and the jet engine will produce electricity and a thrust force.

As the process of producing hydrogen and oxygen gas of course the Water Molecule Power Reactor System will require an Initial Reaction Energy. This Initial Reaction Energy will be supplied by a Solar Power System as based on the previous studies. The Initial Reaction Energy is an energy which will be required so that the electrolysis process in the Water Molecule Power Reactor System could kick start. However once the electrolysis process have started the Initial Reaction Energy will now be minimized since the system will now power on its own given that the Jet Engine System is now producing electricity to power the Water Molecule Power Reactor System.

Sound a bit confusing this book will illustrate the above mentioned principles in a simplified block diagram to illustrate in a simple way as possible what the author wishes to convey to the reader as shown below.

Figure 1. Energy Flow Simplified Diagram

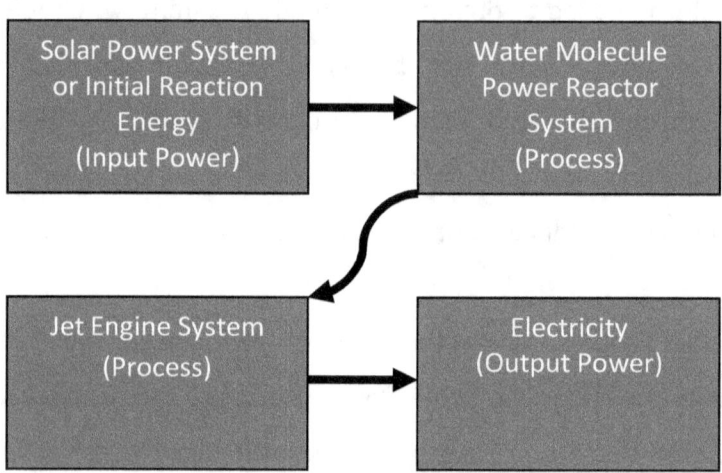

In Figure 1. The electricity output power will be produced by the jet engine system and once the electricity output power is produced it will give back a percentage of its electricity produced to the Water Molecule Power Reactor System and the Solar Power System thereby augmenting the energy produced by the solar power system.

When the energy threshold requirement of the solar power is achieved as provided by the Output Electricity produced then the Water Molecule Power Reactor System will now be independent of the Solar Power System but will now be dependent on the Output Electricity as produced by the Jet Engine.

Simply means to say that the solar power system will only be an introductory power supply to the water molecule power reactor system while the jet engine will be the one to fully power the water molecule power reactor system also providing the lost energy of the solar power upon the initial power consumption of the water molecule power reactor system.

In this process the cycle of the whole system is sustainable as long as the Water Molecule Power Reactor System is producing hydrogen and oxygen gas to the Jet Engine System and the Jet Engine is producing enough electricity.

In this system the solar power will produced Direct Current (DC) to the Water Molecule Power Reactor System and the Jet Engine System is expected produce Alternating Current (AC).

Therefore the power produced at the output is far much more; greater than the energy required for the whole system to operate as it is the main priority objective of this study.

Simply to say that it is the main task of this study to produce a significant amount of electricity at the output which is much greater than the energy required to kick start the energy production of the whole system.

Now at this point in time it will be prudent and wise to look into the characteristics of a jet engine system starting with the most significant subject of Air and Fuel ratio.

To begin with the discussion this book will examine the Air Fuel Ratio first.

Air and Fuel Ratio is defined in reference to Wikipedia as being the mass of air to a solid, liquid, or gaseous fuel present in a combustion process.

Also according to Wikipedia Air Fuel ratio is the ratio between the mass of air and the mass of fuel in the fuel air mix at any given moment.

Before we begin to mathematically define Air and Fuel Ratio this paper will first define Fuel Air Ratio (FAR).

$$FAR = 1 / AFR; \qquad (1.0)$$

Where:

FAR = Fuel Air Ratio;

AFR = Air Fuel Ratio;

By simple understanding Fuel to Air Ratio is the ratio of the quantity of Fuel and Air inside the Jet Engine at the exact moment of internal combustion.

In this paper the fuel being considered is in the form of gas which is Hydrogen and Oxygen gas in perfect mixture since they came directly from sea water.

Other gasses are present of course if sea water is subjected to electrolysis, and other gasses will be filtered by natural compound such as charcoal and others as mentioned in the earlier versions of this paper.

Now again this paper will now define the molecular properties of the hydrogen and oxygen atoms as being relevant to the study at hand.

The basic atomic structure of water is hydrogen and oxygen which is two atoms of hydrogen and one atom of oxygen which was perfectly combined through the act of natural evolution.

To look at the atomic properties of hydrogen and oxygen this paper will examine the following table.

Table 1. Percentage Composition by Element

Elements	Symbol	Atomic Mass	Number of Atoms	Mass Percentage
Hydrogen	H	1.00794	2	11.190 %
Oxygen	O	15.994	1	88.810 %

The table above shows the atomic mass of hydrogen and oxygen, which is in the table it shows that hydrogen composition is only 11.190% while oxygen is 88.810% in a single molecule of water.

In Table. 1 it shows the atomic properties of the single atom of hydrogen and oxygen respectively. At this point in time this paper will now examine the atomic and molecular properties of a single molecule of water. The atomic property of a single molecule of water will be shown in the following table below.

Table 2. Atomic Composition by Element with Water

Elements or Molecule	Symbol	Atomic Mass	Number of Atoms	Number of Molecule
Water	H2O	18	3	1

Now the values mentioned in Table 1 and Table 2 will be essential to get the value of Air and Fuel Ratio which is what this book is trying to clarify.

Now going back to the formula of Air and Fuel Ratio, it was shown as follows.

FAR = 1 / AFR; (1.0)

Where:

FAR = Fuel Air Ratio;

AFR = Air Fuel Ratio;

Now another formula is the Fuel Air Equivalence Ratio also designated by the symbol of phi (Ø). The Fuel Air Equivalence Ratio is defined mathematically as shown.

(Ø)   = (fuel-to-oxidizer ratio) / (fuel-to-oxidizer ration)st;

(Ø)   = [m(fuel)/m(ox)] / [m(fuel)/m(ox)]st;

(Ø)   = [n(fuel)/n(ox)] / [n(fuel)/n(ox)]st;   (2.0)

Where,

[m(fuel)/m(ox)] = is the mass of fuel divided by the mass of oxygen in terms of atomic mass.

n(fuel)/n(ox) = is the number of moles of the fuel divided the number of moles of the oxygen or oxidizer.

Oxygen is considered to be an essential component in combustion and is also dominant in percentage composition in the air being 20 % in comparison to other atomic elements present in the air, the rest of the atomic elements present in the air is less than 20 %.

However in this paper since hydrogen and oxygen is the main fuel for the jet engine and air will only augment in the combustion process in a minimal percentage required in the jet engine systems.

The presence of air in the hydrogen and oxygen combustion will boast the combustion process but the existence of hydrogen and oxygen is already enough to create a combustion process in the jet engine.

To proceed with the computation of the Fuel Air Ratio (FAR) this paper will examine the following formula as shown below. m(fuel)/m(ox) = is the mass of fuel divided by the mass of oxygen or oxidizer in terms of atomic mass.

Simplifying as shown,

m(fuel) / m(ox);

where,

m(fuel) = 18 grams/mol for water which is (H_2O) in gas form;

m(ox) = 32 grams/mol for oxygen in the air (O_2) in gas form;

or in simplified form it can be shown below.

m(fuel) = 18 for water which is (H_2O) in gas form;

m(ox) = 32 for oxygen in the air (O_2) in gas form;

Thus substituting to the formula, it will be shown as follows.

m(fuel) / m(ox) = [18] / [32];

m(fuel) / m(ox) = 0.5625;

Therefore the result of the computation based on the values of the mass of the fuel and mass of the oxygen or oxidizer in terms of mass of the atoms or atomic mass is m(fuel) / m(ox) = 0.5652;

Another value which this book will calculate is the value of the number of molecules of the fuel and the number of molecule of the oxygen or oxidizer which is written in the following formula as mentioned earlier in the discussion.

n(fuel)/n(ox) = is the number of moles of the fuel divided the number of moles of the oxygen or oxidizer.

Since in the above definition it mentions the number of moles of the fuel and the number of moles of the oxygen, it will be prudent of course to illustrate in a table form the number of moles of water and the number of moles of oxygen.

The following table the number of moles of oxygen in the air illustrated and the number of mole of water is also clearly defined as shown.

Table 3. Molecular Composition of Water and Oxygen

Elements or Molecule	Symbol	Atomic Mass	Number of Atoms	Number of Molecule
Water	H2O	18	3	1
Oxygen	O2	32	2	2

Oxygen in air is equal to O2 or two atoms of oxygen.

Thus to calculate for the quantity as mentioned in the formula below and in reference to Table 3, it follows as shown.

n(fuel)/n(ox) = is the number of moles of the fuel divided the number of moles of the oxygen or oxidizer.

n(fuel)/n(ox) =   [1] / [2];

n(fuel)/n(ox) =   0.5;

Thus this paper has now been able to calculate the values needed from the given formula.

$$(\emptyset) = [n(fuel)/n(ox)] / [n(fuel)/n(ox)]st; \qquad (2.0)$$

Where, $(\emptyset)$ is the formula for the Fuel Air Equivalence Ratio.

Thus substituting the derived quantities.

$$m(fuel) / m(ox) = 0.5625;$$

$$n(fuel)/n(ox) = 0.5;$$

From the formula of Fuel Air Equivalence Ratio it can be shown as follows.

$$(\emptyset) = [n(fuel)/n(ox)] / [n(fuel)/n(ox)]st; \qquad (2.0)$$

$$(\emptyset) = [m(fuel)/m(ox)] / [m(fuel)/m(ox)]st; \qquad (2.1)$$

Where,

$$m(fuel) / m(ox) = 0.5625;$$

$$n(fuel)/n(ox) = 0.5;$$

then,

(Ø)   = [m(fuel)/m(ox)] / [m(fuel)/m(ox)]st;   (2.1)

Now if this paper would assume the following.

[m(fuel)/m(ox)]st   = n(fuel)/n(ox) =   0.5;

And,

m(fuel) / m(ox) = 0.5625;

Therefore,

(Ø)   = [m(fuel)/m(ox)] / [m(fuel)/m(ox)]st;   (2.1)

Given that,

[m(fuel) / m(ox)]   = 0.5625;

[m(fuel) / m(ox)]st   = 0.5;

Substituting the values to the formula in the following computation as shown.

(Ø)   = [m(fuel)/m(ox)] / [m(fuel)/m(ox)]st;   (2.1)

$(\emptyset) = [0.5625] / [0.5]\text{st}$;

$(\emptyset) = 1.125$;

Based on the above computation it is shown that the Fuel Air Equivalence Ratio is $(\emptyset) = 1.125$;

It is stated that Fuel Air Equivalence Ratio that is greater than one always mean that there is more fuel in the fuel oxidizer mixture than required for complete combustion.

Thus from the derived value of Fuel Air Equivalence Ratio which is $(\emptyset) = 1.125$, then this value will give more fuel for good combustion inside the jet engine.

Also from the formula of Fuel Air Ratio (FAR) and Air Fuel Ratio (AFR) which is stated mathematically as shown.

FAR = 1 / AFR;   (3.0)

Also,

$\lambda = [\text{AFR}] / [\text{AFR}]\text{stoich}$;

where, (λ) is the Air Fuel Equivalence Ratio.

Now if this paper will assume that (λ) = AFR,

Then,
FAR = 1 / AFR; (3.0)

If this paper would let, (λ) = AFR, then.

FAR = 1 / AFR;   if (λ) = AFR,

Thus,

FAR = [1 / λ];

FAR = 1 / λ; (3.1)

Also if this paper will let FAR = ($\emptyset$), then it can written as follows.

FAR = 1 / λ;   if FAR = ($\emptyset$),

Thus,

$\emptyset$ = 1 / λ; (3.2)

And,

$$\lambda = 1 / \emptyset; \quad (3.2.1)$$

Since from the previous computation the value of $(\emptyset) = 1.125$ has been defined. Thus this paper will substitute the value of $(\emptyset) = 1.125$ to the following formula.

$$\lambda = 1 / \emptyset; \quad (3.2.1)$$

If $(\emptyset) = 1.125$, then it follows.

$$\lambda = [1 / \emptyset]; \quad (3.2.1)$$

$$= [1 / 1.125];$$

$$\lambda = [0.889];$$

Thus the value of the Air Fuel Equivalence Ratio ($\lambda = 0.889$), as shown in the computation.

It is stated that ($\lambda < 1.0$) is rich mixture which is in the derive value of $\lambda = 0.889$, this value is equal to $\lambda = 0.9$ or is lesser than 1.0 therefore the mixture is rich.

For rich mixture the power of the jet engine will increase and will produce more heat while fuel consumption will be lesser.

Thus now this paper had now been able to establish the principles of Air Fuel Ratio and Fuel Air Ratio which is very essential in the analysis of the jet engine propulsion system.

At this point this paper will summarize the properties of oxygen as being a component of air.

Table 4. Properties of Oxygen in the Air

Elements or Molecule	Symbol	Mol/Mol air	Vol. %	Molar Mass	g/ Mol air	wt %
Oxygen	O2	0.20946	20.446	32	6.702469	23.14

Based on available data oxygen is 20.446 percent in relation to other gaseous elements present in air. Meaning oxygen is more dominant in air but is not the only gaseous element present in air.

Also the molar mass of air or dry air is computed to be 28.97g/mo.

Review Questions

1. Define the meaning of Air Fuel Ratio.
2. Define the meaning of Fuel Air Ratio.
3. What is the difference of Air Fuel Ratio Equivalence and Air Fuel Ratio?
4. What is the difference of Fuel Air Ratio Equivalence and Fuel Air Ratio?
5. What is the significance of studying Fuel Air Ratio and Air Fuel Ratio?

Problem Solving

1. What is the atomic mass of hydrogen and oxygen as separate atoms respectively?
2. What is the atomic mass of hydrogen and oxygen as being a single molecule of water?
3. What is the mass percentage of hydrogen and oxygen respectively inside the molecule of water?
4. Calculate the Fuel Air Ratio Equivalence if assuming the Air Fuel Ratio Equivalence is equal to 1.565.
5. Why is oxygen considered as the main oxidizer in this study?

CHAPTER TWO
Performance Efficiency Factor

Water Molecule Power Reactor System

> *"*
> *This chapter will examine the fundamental mathematical principles of the performance efficiency factor with the use of the Water Molecule Power Reactor System.*

Analysis of the electric generation capacity and efficiency of the proposed Water Molecule Power Reactor System

In the previous edition of this book some technical details on how the proposed Water Molecule Power Reactor System will be able to produce electricity where explained both mathematically and in qualitative point of view.

In this edition other quantitative values of great interest will also be added to shed light on some technical details concerning on how would the proposed water molecule power reactor system would be able to produce electricity as it was hoped in this study.

To begin with this task, this would examine the variable of device efficiency or how efficient would the proposed reactor be in terms of electric power generation.

This book will now examine the device efficiency of the proposed reactor with the help of the following formula as shown below.

Device efficiency = Useful energy output / Energy input;

From the previous study it was measured and shown mathematically some variables which are important in the input power, process power and output power as can be shown with the following table summary.

Table 5. Power Generated Summary

Input Solar Power (watts) DC	Process Force / Thrust Generated (Newton)	Jet Engine Output Power (watts) AC
600 watts	2,075.5 km.kg/ s^2	2.6 kilowatts

Now based on the data from the previous edition of this book and as reflected on the summarized table in Table 5, this book will now compute for the device efficiency as follows.

Device efficiency = Useful energy output / Energy input;

If this paper will denote Device Efficiency as DE, Useful Energy Output as UEO and Energy Input as EI, then the formula can be written as.

$$DE = (UEO / EI) \times 100\%; \qquad (4.0)$$

$$DE = [(2.6 \text{ kilowatts}) / (600 \text{ watts})] \times 100\%;$$

$$DE = 4.3 \times 100\%;$$

$$DE = 433\ \%;$$

Based on the result of the computation it states that Device Efficiency is DE = 4.3, meaning that the Output Power generated is 4.3 times more than what was the Input Power used in the process and that the DE = 433% or 400 times compared to the input power of 100 percent.

In terms of power efficiency the power generated is 400 times greater than the input power. Thus therefore the system is super efficient.

Another variable to consider in relation to device efficiency is the Coefficient of Performance or COP which define on how well the device perform in terms of output efficiency and can be mathematically expressed as follows.

COP = (UEO / EI); (5.0)

And based on Table 5, this paper can compute the COP by the values and quantities presented in Table 5, as shown below.

COP = (UEO / EI); (5.0)

COP = (2.6 kilowatts) / (600 watts);

COP = 4.3;

Based on the value of COP = 4.3, this means that the performance factor is greater than 1, which is what is minimally required for the system to be efficient in terms of performance and the value of 4.3 is four times than the minimum value of 1, thus therefore the performance in terms of electric power generation is very good.

It is also good to know that the input power of 600 watts is produced by the solar panel which generates a Direct Current (DC) voltage, while at the output of the jet engine the output power generated is 2600 kilowatts Alternating Current (AC).

The AC power generated at the output will be subdivided by the system as its internal source of energy for self sufficiency capability of the reactor itself.

Review Questions

1. Why is it important to understand the efficiency factor the proposed water molecule power reactor system?
2. Based on the computations shown the output power was greater than the input power. What will happen if the input power is much greater than the output power?
3. What is the difference between an alternating current and a direct current?
4. Can the alternating current and power produced at the output be converted back to direct current for use in the internal power requirement of the proposed reactor?

5. What is the meaning of the COP value if it is lesser than 1, and what is the meaning if it is greater than 1?

Problem Solving

1. If the COP is 1.5 and the EI is 300 watts AC, calculate the value of the UEO.
2. Calculate the value of DE if UEO is 600 AC watts and the EI is 600 DC watts. Is the system efficient in terms of power generation?
3. If DE less than 1 what does it mean, if DE is equal to 1 what does it imply and finally if DE is greater than 1 what does it mean for a given system in terms of its performance to deliver productive output.
4. Calculate the value of DE if UEO is 1200 DC watts and the EI is 1000 DC watts. Is the system efficient in terms of power generation?
5. If the COP is 0.5 and the EI is 800 watts AC, calculate the value of the UEO.

CHAPTER THREE
Jet Engine Electricity Generation Capacity

> *This chapter will examine the fundamental mathematical principles of the Jet Engine and the ways it is able to generate electricity with the use of the Water Molecule Power Reactor System.*

Analysis of the electric generation capacity of the proposed Water Molecule Power Reactor System

At this point in time this book will now examine the electric generation capability of the proposed reactor and it is within the aim of this writing that the proposed reactor will produce much energy than what is consumed in the initial start of its reaction process.

One given example will be focused on how would the reactor produce electricity by the given quantity of thrust force as previously mentioned and analyzed.

Now the thrust force was computed to be equal to Thrust = 2,075.5 kg.km/s², this thrust force will cover a distance of 2.075 km/s or 7.47 km/hour at a given moment in time.

TF = thrust force;

TF = $2,075.5 \times 10^3$ kg.m/s²;

D = distance;

D = 2.075 km/s; or D = 7.47 km/hour;

Now if we would use the formula for the Power Generation capacity for the wind turbine to see if the Thrust Force generated by the proposed reactor can in fact produce electricity if it would power up a wind turbine.

From the formula,

$P = \frac{1}{2} \rho A v^3 C_p;$ 6.0

Where,

$L = r$ = Blade Length;

V = Wind Speed;

ρ = Air Density;

Cp = Power Coefficient;

$A = 2 \pi r^2$;

Now if we would let the following values be defined, $L = 0.333$m, V = Thrust Force, $\rho = 1.23$ kg/m³, $Cp = 0.4$; with V = Thrust Force = 2,075.5 kg.km/s²;

$$P = \tfrac{1}{2} \rho A v^3 Cp; \qquad\qquad 6.0$$

$= (1/2)\ (1.23\ \text{kg/m}^3)\ (2\pi\ 0.333^2)\ (2{,}075.5\ \text{kg.km/s}^2)^3\ (0.4)$;

$= (1/2)\ (1.23\ \text{kg/m}^3)\ (0.697)\ (8.941 \times 10^9\ \text{kg.km/s}^6)\ (0.4)$;

$P = 1.533 \times 10^9$ kilowatts;

$P = 1.533 \times 10^9$ kilowatts/second;

Or equivalently,

$P = 1,533 \times 10^6$ Megawatts/second;

Wow that is also a very large amount of power to be generated in one second of a moment in time.

Therefore this book can say that the power generated by the Water Molecule Power Reactor System is far more than it can consume in a given time of one second.

The power requirement of the miniature scale is P = 306.48 watts DC;

The power requirement for the industrial scale is Power = 423.564 Watts;

Even the sum of the two initial power requirements is just a tiny fraction of the power generated by the cylindrical reactor.

Wind Turbine is very similar in power generation analysis in comparison to jet engine. Jet engines are even more efficient in terms of power output compared to wind turbines.

Therefore if the cylindrical reactor is integrated to jet engine too much power will be produced significantly.

There are numerous jet engine designs that will fit the technical specifications of the proposed cylindrical reactor and surely the electric power generation capability will be very efficient, effective and enormous.

Review Questions

The following questions will help the reader understand the topics mentioned in this chapter

1. Why is it important to analyze the thrust force in relation to the development of the water molecule power reactor system?
2. What is the significant relationship of the thrust force and the distance travelled by a given force?
3. What is the difference by the thrust force generated within a given second of time and the elapsed time of one hour?
4. What is the meaning of a gross thrust force and a net thrust force?
5. What is the importance and essence of a thrust force in relation to generating electricity?
6. What characteristic of the wind turbine which are similar to that of the jet engine?
7. Based on the computation as shown in this chapter and assuming that thrust is constant. Is the power of the wind turbine constant or it varies with the speed of the wind

8. What is the reason behind in using the wind turbine as an example of how the proposed reactor would be able to generate electricity?
9. What is the difference between the jet engine and wind turbine in terms of how it would produce electricity?
10. In the example as shown in this chapter is the wind turbine dependent on the speed of the wind or dependent on the velocity of the proposed reactor?

Problem Solving

1. What is RMS velocity value of hydrogen and oxygen respective?
2. Calculate the RMS velocity of H_2O and then compute the thrust force if the mass of the object being propelled is 2000 grams.
3. From the above result on the power generated by the wind turbine, calculate the net power generated if 20% of the power generated is used for the power igniter in the cylindrical reactor.
4. Calculate the power generated by the wind turbine given that all values are the same as previously computed except if the length of the blade is reduced to 0.2 meter.

5. Calculate the power generated by the wind turbine given that all values are the same as previously computed except if the length of the blade is 0.2 meter and the Power Coefficient is reduced to 0.05 respectively.

CHAPTER FOUR

Electron Quantity Requirements for the Water Molecule Power Reactor System Design

> *This chapter will examine the fundamental mathematical principles of the electron quantity requirements for a given, specific design parameters of the Water Molecule Power Reactor System.*

Computation for the electron quantity with the Miniature Scale Model of the Cylindrical Reactor System

Now this book will now compute for the miniature scale design of the water molecule power reactor system as will be shown below.

Computing for the area of the cylinder for the miniature scale model of the cylindrical reactor,

With $L = 6$ inches, $r =$ inch;

Recalling the formulas as follows,

$$A = (2\pi r^2) + h(2\pi r); \qquad 2.0$$

And

$$R = [\rho L / A] \qquad 3.0$$

Computing for the area of the cylindrical reactor with L = 6 inches, r = 2 inches, it follows as shown.

$$A(\text{cylinder}) = (2\pi r^2) + h(2\pi r); \qquad 2.0$$

$$= (2\pi)(2^2) + 6(2\pi)(2);$$

$$= (25.133) + (75.40);$$

$$A(\text{cylinder}) = 100.531 \text{ inches}^2;$$

Computing for the volume of the cylinder as follows,

$$\text{Volume of cylinder} = (2\pi r h);$$

$$= (2 \pi r h);$$
$$= (2 \pi\ 2\ 6);$$

Volume of cylinder = 75.40 inches³;

Now this book will compare the volume of the proposed reactor to the equivalent volume of 1 liter to see how many liters are there present in this proposed cylindrical reactor.

1 liter = 1000 cm³; now 1 inch = 2.54 cm;

(2.54 cm)³ = 16.38 cm³;

Then we can say that,

1 liter = (1000 cm³) x (1 inch³/16.38 cm³) = 61.050 inch³;

Therefore 1 liter = 61.05 inch³,

Using the following formula as shown,

Volume of the Fuel = Volume of the Reactor / 1 liter Volume;

$$= (75.40 \text{ inches}^3) / (61.05 \text{ inch}^3);$$

Volume of the Fuel = 1.2 liters;

Therefore with the Length = 6 inches and radius = 2 inches the volume of sea water that can be used as fuel for the cylindrical reactor is equal to 1.2 liters maximum but of course the fuel that can be used in this design will be only 1 liter minimum

One liter is the standard of measure being adopted in this research study and in this book.

Computing for the Resistance value of the cylindrical reactor with sea water as fuel inside the reactor. With the resistivity constant value of sea water equal to ρ = 0.2 Ω. x (39.36 inches) = 7.872 Ω.inches;

R(fuel) = [ρL/ A]   3.0

$$= [(7.872 \ \Omega.\text{inches}) (6 \text{ inch})] / [100.531 \text{ inches}];$$

R(fuel) = 469.825x10$^{-3}\Omega$;

Computing for the area of the rods with L = 5 inch and r = (2.5 mm) x (1 inch / 25.4 mm) = 0.98 inch;

$$A(rods) = (2 \pi r^2) + h (2\pi r); \qquad 2.0$$

$$= (2 \pi) (0.98^2) + 5 (2\pi) (0.98);$$

$$= (6.034 \text{ in}^2) + (30.79 \text{ in}^2);$$

$$A(rods) = 36.822 \text{ inches}^2;$$

Computing for the Resistance of the Rods with L = 5 inches, A(rods)= 36.822 inches² and ρ= 2.715 x 10^{-5} Ω.inch; in inches value for the stainless steel material.

Computing as shown below,

$$R(rods) = [\rho L / A] \qquad 3.0$$

$$= [(2.715 \times 10^{-5} \, \Omega.\text{in}) (5 \text{ in})] / [36.822 \text{ in}^2];$$

$$R(rods) = 3.687 \times 10^{-6} \, \Omega;$$

Computing for the total resistance of the rods inside the cylindrical reactor with a total of eight rods,

R(rods total) = R(rods) x 8;

$= 3.687 \times 10^{-6} \Omega \times 8$;

R(rods total) = 29.493×10^{-6} ohms (Ω);

Computing for the Total Overall Resistance of the Cylindrical Reactor with R(fuel) = $469.825 \times 10^{-3} \Omega$ and R(rods total) = 29.493×10^{-6} ohms (Ω);

RT(total) = R(fuel) + R(rods total);   4.0

RT(total) = R(fuel) + R(rods total);

$= (469.825 \times 10^{-3} \Omega) + (29.493 \times 10^{-6}$ ohms, $\Omega)$;

RT(total) = $469.854 \times 10^{-3} \Omega$;

Computing for the overall current requirement of the proposed miniature scale model cylindrical reactor, with V = 12 volts DC and RT(total) = $469.854 \times 10^{-3} \Omega$;

I = V / R;

= (12 volts DC) / ($469.854 \times 10^{-3} \Omega$)

I = 25.54 amperes DC;

DC for direct current.

Computing for the power requirement of the system, it follows as shown,

P = I V;

= (25.54 amps) x (12 volts DC);

P = 306.48 watts DC;

DC for direct current.

At this point in time this book had now technically been able to compute how much current will be required by the cylindrical reactor in a given time of one second and the power needed by the reactor also in one second.

The next computation that this book will now examine is how much electricity that the reactor can produce in one second so that the needed current to operate the reactor must be lesser that what quantity of current the reactor can produce in a given time of one second.

From the previous edition of this book the quantity of electrons per one ampere of current was quantified and was mathematically shown to be as shown below.

From,

1 A = 1 C / 1 T (second);

Therefore if the current value is equal to 25.54 A, then this computation can be shown as follows, according to Physics,

1 coulomb = 6.25×10^{18} electrons;

And,

I (current) = 25.54 A;

Then this paper shall convert this value of current to the corresponding number of electrons in a given current quantity; this paper shall use the following conversion process,

I (amperes) = I (given current, A) x $(6.25 \times 10^{18}$ electrons/ 1A);

= (25.54A) x $(6.25 \times 10^{18}$ electrons/ 1A);

I (amperes) = 1.596×10^{20} electrons/second;

Now this paper will also compute for the voltage in terms of electron quantities in volts or simply designated as follows, V(e-).

V(e-) = [I(e-)] x [R]; (5.0)

Where, RT(total) = $469.854 \times 10^{-3} \Omega$, and I (amperes) = 1.596×10^{20} electrons/second, respectively.

Substituting values as shown below.

$$V(e-) = [I(e-)] \times [R];$$

$$= [1.596 \times 10^{20} \text{ e-/s}] \times [469.854 \times 10^{-3} \Omega];$$

$$V(e-) = 7.50 \times 10^{19} \text{ electron volts,}$$

The above value represents the quantity of electrons required inside the reactor in a given time of one second.

Now this paper will calculate the time required before the one liter of water inside the reactor to fully deplete.

Let this paper establish the following formula as used in the previous methods of the previous studies conducted by the author and previous edition of this book.

$$w(t) = dQ / dT; \qquad (6.0)$$

Where:

$w(t)$ = a value (H2O) which is a function of time;

dQ = a value (H2O) which is a measure of quantity;

dT = a value which is a measure of quantity with respect to time;

Now let, dQ = N(e⁻)H2O, and let dT = V(e⁻), and so,

$$w(t) = dQ / dT; \qquad (6.0)$$

$$= [N(e^-)H2O] / [V(e^-)];$$

$$w(t) = [N(e^-)H2O] / [V(e^-)]; \qquad (6.1)$$

Where $N(e^-)H2O = 1.0038 \times 10^{27}$ electrons as shown in the previous edition of this book.

The values that appear on $N(e^-)H2O = 1.0038 \times 10^{27}$ electrons, is the quantity of electrons present in one molecule of water.

Since $V(e-) = 7.50 \times 10^{19}$ electron volts and $N(e^-)H2O = 1.0038 \times 10^{27}$ electrons, is given then w(t) can be found as shown below.

$w(t) = [N(e^-)H2O] / [V(e^-)]$; (6.1)

$= [1.0038 \times 10^{27} (e-)] / [7.50 \times 10^{19} (e-)]$;

$w(t) = 13.384 \times 10^6$ molecules of H2O per hour consumption.

The above value of w(t) represents the quantity of water molecule required in the proposed reactor to be able to produce 1000 grams of combined hydrogen and oxygen gas in a given one second of time.

To see on how many hours will the quantity one 1000 grams or 1 liter of water will be consumed in a given time. This paper will calculate the following.

From,

$dQ = w(t) \times dT$; (6.2)

where, $w(t) = 13.384 \times 10^6$ molecule of H2O and dT = 1 hour/3600 seconds.

$dQ = w(t) \times dT;$ (6.2)

Substituting the values as shown,

$dQ = w(t) \times dT;$ (6.2)

$= [13.384 \times 10^6 \text{ molecules}] \times [1 \text{ hour}/3600s];$

$dQ = 3.717 \times 10^3$ molecule hours before the water is fully depleted inside the proposed reactor.

Now from the chemical reaction in the cathode and anode the statement below shows that it requires 2 electrons in the cathode to produce hydrogen and 2 electrons in the anode to produce oxygen.

Cathode (reduction): $2 H_2O(l) + 2e^- \rightarrow H_2(g) + 2 OH^-(aq)$

Anode (oxidation): $2 OH^-(aq) \rightarrow 1/2 O_2(g) + H_2O(l) + 2 e^-$

Therefore the electrolysis process requires a total of 4 electrons both in the anode and the cathode terminals.

From the computed values,

$dQ = 3.717 \times 10^3$ molecule hours before the water is fully depleted inside the proposed reactor.

Let dQ be divided by the 4 electrons as shown.

$dQ = [3.717 \times 10^3$ molecule hours$] / [4$ electrons$]$;

$dQ = 929.25$ hours;

Now the value of dQ represents the quantity of water inside the proposed reactor after one hour of electrolysis process.

Now this paper will divide the dQ with 24 hours in a day. Thus now,

$dQ = 929.25$ hours;

$= [929.25$ hours$] / [24$ hours per day$]$;

dQ = 38 days;

Now the value of dQ has been converted to days, which means that the water inside the proposed reactor with the given dimensions and water content will last for 38 days.

Now 38 days is one month and seven days.

Review Questions

The following questions will help the reader understand the topics mentioned in this chapter

1. What is the substance resistance value of sea water?
2. What is the meaning of substance resistance value?
3. What is the substance resistance value of stainless steel?
4. What is the primary reason why the stainless steel is used for the conductor rods in the proposed reactor?
5. Why use a cylindrical shaped reactor? What is the essence of this design?
6. Why is it important to be able to determine the resistance of a given material or substance in terms of power generation?
7. Are there other materials which can be used for the conductor rods in the proposed reactor?
8. Why it is that there are eight conductor rods in the reactor?
9. If the conductor rods are increased to twelve pieces will the resistance of the reactor cylinder change?
10. Does the dimension of the cylinder reactor affect the overall power generating capacity of the proposed reactor?

Problem Solving

1. Calculate the resistance value of cylinder if the length is 20 inches and radius 3 inches using sea water as its fuel.
2. Calculate the resistance value of a given cylinder with length is equal to 6 inches and radius is 2 inches using sea water as its fuel with rods length of 5 inches and radius 0.98 inches, the number of rods in the reactor is 12 pieces with rods used is stainless steel rods.
3. Based on the discussion and computation in the previous chapter and in this chapter what is the value of the industrial scale reactor total resistance and the miniature scale reactor total resistance, find their ratio?
4. Calculate the resistance value of a given cylinder with length is equal to 6 inches and radius is 2 inches using sea water as its fuel with rods length of 5 inches and radius 0.45 inches, the number of rods in the reactor is 12 pieces with rods used is stainless steel rods.
5. Calculate the resistance value of a given cylinder with length is equal to 6 inches and radius is 2 inches using sea water as its fuel with rods length of 5 inches and radius 0.45 inches, the number of rods in the reactor is 12 pieces with rods used is silver material rods.

CHAPTER FIVE
Analysis of the energy or force generated by the Water Molecule Power Reactor System

> *This chapter will examine the fundamental mathematical principles of the force or energy produced by the proposed Water Molecule Power Reactor System.*

Analysis of the energy or force generated by the Water Molecule Power Reactor System

Thus moving on with the discussion, out attention will now be moved to the next concern of this book which is the power output of the system which of course had already been discussed thoroughly in the past edition of this book but will however be discussed again in this special edition. The discussion will however be expounded more specifically for the readers much understanding on the subject.

Table 2. Force, Thrust values for the sea water at 1 liter volume

Sea Water Volume Input Fuel	1 liter to kilogram Value Output	Thrust Force Output or velocity	Fuel Life Cycle for 1 liter volume	Distance to be covered for 1 liter
1 liter of sea water	1000 grams of pure H2O gas	2,075.5 kg.km/s²	1.032 hour	2.075 km/s or 7.47x10³ km/hr

Table 2, explains that for 1 liter of sea water inside the proposed water molecule power reactor system it will produce and equivalent of 1000 grams pure hydrogen and oxygen gas combined.

Also from Table 2, it also explains that for 1000 grams of H2O gas it will produce a thrust force of 2,075.5 kg.km/s² within a 1.032 hour of electrolysis process inside the water molecule power reactor system. The thrust force of 2,075.5 kg.km/s² can reach a distance of 2.075 km/s or 7.47 x10³ km/hr.

The value of distance and thrust force is vital for the analysis of how will the jet engine produce power given the value of thrust force which is in this values it is very promising. The thrust force will of course be affected by the weight of the object being propelled into the sky but of course the thrust force is high enough to transport 1kg of weight into the sky.

The value of thrust force divided by the weight of the object can be calculated as follows.

$$TF(net) = TFO(gross) / WO; \quad\quad\quad 1.0$$

Where:

$TF(net)$ = Net Thrust Force;

$TFO(gross)$ = Gross Thrust Output Value;

WO = Weight of the Object;

Also in other form it follows as shown below.

Net Thrust Force = Gross Thrust Output Value / Weight of the Object;

To give an example we will calculate the net thrust force as shown.

$$TF(net) = TFO(gross) / WO; \quad\quad\quad 1.0$$

The problem will be as follows how much net thrust force will be generated by the water molecule power reactor system if the

Gross Thrust Force Output = 2,075.5 kg.km/s² and the Weigh of the Object being Propelled is 1000 grams.

Solution:

Given,

Gross Thrust Force Output = 2,075.5 kg.km/s²

Weigh of the Object being propelled is 1000 grams.

Find,

Net Thrust Force = ?

Solution,

From the formula,

TF(net) = TFO(gross) / WO;   1.1

Then,

TF(net) = TFO(gross) / WO;

= (2,075.5 kg.km/s²) / (1000 grams);

TF(net)= 2,075.5 km/s²;

From physics,

1 Newton (N) = 1 kg.m/s²;

Thus,

TF(net)= 2,075.5 kg.m/s²;
       = 2,075.5 kg.m/s²;
TF(net)= 2,075.5 kg.m/s²;

Or equivalently,

TF(net)= 2,075.5 Newton or TF(net) = 2.0755 Kilo Newton;

Or simply TF(net)= 2,075.5 kg.m/s²;

This means that a 1 kilogram (kg) of object can be accelerated to a velocity of 2,075.5 m per second per second.

Now 1 hour is 3600 seconds. So multiplying this values to the TF(net) of 2,075.5 kg.m/s². We have as follows.

TF(net)= 2,075.5 kg.m/s²;

$= (2{,}075.5 \text{ kg.m/s}^2) \times (3600 \text{ seconds})$;

TF(net)= $7{,}471.8 \times 10^3$ kg.km/second;

This means that 1 kg of object can be accelerated to $7{,}471 \times 10^3$ kilometers per second in a given moment of time for one second only.

To convert the value to one hour to better understand the idea it follows as shown.

TF(net)= $7{,}471.8 \times 10^3$ kg.km/seconds;

$= (7{,}471.8 \times 10^3 \text{ kg.km/seconds}) \times (3600 \text{ seconds/1 hour})$;

TF(net)= 26.898×10^6 kg.km/hour;

This means that after one hour the object of 1 kilogram is already 26.898×10^9 meters away from where it was in its initial position.

Therefore by this computation the thrust force is so great for only 1 kilogram of object being propelled, such a value can even lift an object with a weight of 1 ton based on the above computation.

The thrust force will be used to accelerate our jet engine and in turn the jet engine will produce AC power out and in this AC power out a fraction of this will be injected back to the Water Molecule Power Reactor Device will only needs a DC power to operate.

In this technical analysis the output power produced is much greater that what is required by the input power requirement and the whole system will sufficiently operate for one hour.

Review Questions

The following questions will help the reader understand the topics mentioned in this chapter

1. What is the relationship of a gross thrust force and that of a net thrust force?
2. Should the gross thrust force be greater than the net thrust force? If so then why?
3. Is thrust force responsible for the movement of any object in the horizontal and vertical acceleration?
4. What is the difference between a horizontal acceleration and vertical acceleration?
5. Define the meaning of newton as a unit of force?
6. As mentioned in Table 2, how much force will be produced by a, 1000 grams of pure hydrogen and oxygen gas?
7. What is the equivalent value of 1 liter to kilograms?
8. What is the formula for the net thrust force?
9. Will the weight of the object being pushed upward affect the thrust force?
10. Which direction will require much thrust force horizontal acceleration or vertical acceleration?

Problem Solving

1. If the Gross Thrust Force is equal to TF(gross) = 2,075.5 kg.m/s² and the weight of the object being propelled upward is 1500 grams. How much of the gross thrust force will remain.
2. If the object being propelled upward is 5000 grams and the Thrust Force is only, 2,075.5kg.m/s², how many times is the weight heavier compared to that of the thrust force? How many times much thrust force will be required?
3. In what ratio and proportion will the thrust force adjust in relation to the weight of the object being propelled? Say the weight of the object is 6000 grams and the thrust force is 2,075.5 Newton.
4. In the problem as mentioned in no. 3 how many liters of sea water will be required for that particular weight of the object as a fuel for the proposed reactor system.
5. Thrust force generated by the reactor system is a thrust force generated by the fuel itself the thrust force generated by the jet engine is not yet added into the equation. Will the thrust force power of the jet engine add to the thrust force generated by the fuel of the reactor?

CHAPTER SIX

Analysis of the Input and Output Power Requirements for the Water Molecule Power Reactor System

> *This chapter will examine the fundamental mathematical principles of the input power requirements and the power output produced by the proposed Water Molecule Power Reactor System.*

Analysis of the Input and Output Power Requirements for the Water Molecule Power Reactor System

And now this book will now move on to discuss the other components of the Water Molecule Power Reactor System which is the analysis of the input power requirement and the output power generated by the system.

As mentioned in the earlier edition of this book the following design will be considered as follows.

The four, solar panel that will be used in this research will be connected in parallel as will be shown in the calculations below.

Solar Panel in Parallel connection:

Voltage is Constant:

Power	= 150 Watts x 4	= 600 Watts
Voltage	= 17.2 Volts	= 17. 2 Volts
Current	= 8.72 Amp x 4	= 34.88 Amps

Therefore in the above computation the voltage of the solar panel is constant at 17.2 Volts (DC), the power has increased four times from 150 Watts to 600 Watts and the current have also increased four times from 8.72 Amperes to 34.88 Amperes.

Amperes above 8.72 Amps is very dangerous to human health, it can induce cardiac arrest to anyone having in contact with such a high flow of electrical current, even in a Direct Current settings.

Even an ampere value of 1 ampere is very dangerous to human health. Therefore it is medically and scientifically advised to use appropriate personal protective equipment's when handling current above 1 ampere and even in dealing with any current in any type and form.

Table 3. Input Power and Output Power Table

Solar Input Power (watts) 150w x 4	Solar Input Voltage (volts)	Solar Input Current (amperes)	Jet Engine Power Output (watts) 650w x 4	Jet Engine Voltage Output (volts) AC
600 Watts DC	17.2 Volts DC	34.88 Amps DC	2.6 kilowatts AC	110 volts AC or 220 volts AC

Table 3, explains that for the input power this system will utilized four 150 watts solar panel which will be used in this system, which will produce a DC voltage of 17.2 volts and a current of 34.88 amps.

The expected output power of the Jet Engine will be 2.6 kilowatts because four 650 watts engine will be used in this system design which totals to 2.6 kilowatts of AC power output with voltage output also of 110 volts AC or 220 Volts AC.

Around 25% of the output power generated will be used back into the input power so that the whole system is self-sustaining and will be operating round the clock without interference except for repairs, which is why, there would be a backup power system which will only operate in the events of repairs.

In this type of system design it is classified as semi industrial scale prototype design. This type of system is typically designed for the purpose of illustration on the intricate details of the industrial scale and miniature scale designs.

The difference between industrial scale design and the miniature scale design is primarily on the output that will be required of the system. However both systems will function nevertheless the same only the output will vary accordingly to the load.

In the industrial scale model the minimum output of the system will be used by a single to multiple homes.

While the miniature scale design model the minimum output of the system will be used by vehicles for land and air type transport and also for jet engine propulsion system for planetary and interplanetary travel.

Miniature scale design model is typically cylindrical type design which is a reactor that is cylindrical in shape with turn 180 degrees lock system for easy replacements. In the miniature scale design model two cylindrical reactors will operate in the system the first cylinder will run until it is depleted then the next cylinder will be operated automatically when the second cylinder operates the first cylinder will be replaced with a fully fuel loaded cylinder and same goes to the second cylinder.

Meaning there is always one cylinder running at any given time. Fuels will be stored in a cylinders where the cylinders itself is the reactors or a main reactor will be operating with two reserved fuel cylinders this would be economical but cylindrical reactors is the best because they can be repaired anytime and the reactor is always well maintained.

However in other rigid applications one main reactor will be used and two cylinder reactors will also be used with two cylinder fuels. Meaning there are three reactors with only one operating at any given time, the other two are reserved.

Sounds good however it will be good to separate the reactors for the system power and the reactor for the propulsion system and so therefore we will add another reactor for the system power generation. So for rigid applications there are two main reactors and two cylindrical reactors for the propulsion system. In this design there are always two reactors running at any given time one for the system power requirement and the other for the propulsion system.

Now for the two application of the Water Molecule Power Reactor System which is the Industrial Scale and Miniature Scale. This book will elaborate the difference of the two as mentioned earlier.

Industrial Scale design of the Water Molecule Power Reactor System is based on permanent non-moving structures meaning the whole system is like a power generation facility and is designed to generate electric power for a single house or a multiple number of household. Or additionally also to power certain manufacturing establishments or office areas as may deemed essential and appropriate based on the design of the system.

Miniature Scale design is a type of a design which is mobile and moving all the time meaning the system design is for rigid applications such as in the transportation in land on in the air or even in space. The miniature design will be integrated to moving vehicles within large geographical area and also within a given air space. Which is why the design for the miniature water molecule power reactor system is a bit complex because of the nature of its application which is mobile and rigid. A book will be written much specific for the application of the water molecule power reactor system in the industrial scale model and in the miniature scale model respectively.

Now the initial power requirement for the input in the reactor is considered a power initiator or in other term igniter which in this system will initiate the process in the reactor, however once in the process is initiated within a period of about just a second the system will become self-reliant and will automatically power its own system while maintaining constant power supply to the igniter power source.

Without a power igniter into the reactor the reactor cannot operate. However with the introduction of the power igniter the reactor will now sustain itself as long as there is fuel to the reactor and the reactor will operate non-stop while interchanging the supply of fuel to two cylinder fuel tubes.

In the event of reactor turn off then again the power igniter will again be used, thus the power igniter will only be used at the event of reactor power shutdown in the time of idle usage or when the reactor is not being used. If the reactor is being constantly used then the power igniter will be technically idle but is constantly being recharged by the reactor itself.

The design of the power igniter will vary in the industrial and miniature scale design and depending upon the needs of the reactor system. Hybrid system will incorporate and integrate many techniques and technological specifications depending on the prospective design output.

The design of the cylindrical reactors will now be discussed in this book as follows.

Review Questions

The following questions will help the reader understand the topics mentioned in this chapter

1. In the topics mentioned in this chapter what is the power in watts output of a single solar panel?
2. In this chapter how many solar panel will be used for the water molecule power reactor system initial input power?
3. How many watts of power if all the combined solar power is used as proposed on this chapter?
4. What is the output voltage of the single panel is as mentioned in this chapter?
5. Why is that solar panel power system, is used as an initial power output as mentioned in this chapter?
6. Does a solar panel need a battery to store its energy?
7. Based on this chapter how many watts of power can a single jet engine produce?
8. Based on this chapter how many watts in total power can the combined jet engine produce as mentioned above?

9. How many times is the input solar power in watts in comparison to the output power in watts of the jet engine?
10. Why do we need a jet engine for our electric generation needs as mentioned in this book?

Problem Solving

1. Calculate the power ratio of the input power and the output power with data in Table 3.
2. Increase the number of jet engine to six jet engines at the out in relation to the data as mentioned in Table 3, and define how many output power in watts will be produced.
3. Given the data in Table 3, how much power will remain if 20% of the jet engine output power is consumed back to the input of the reactor?
4. If the design is one solar battery per one solar panel how many solar batteries will be required for the system as mentioned in Table 3.
5. If the single solar panel cost $ 80.00 each and the battery is $ 160.00 each how much would it cost if four, solar panel is used with one battery per solar panel.

Glossary

Water Molecule Power Reactor is a cylindrical shaped small reactor module with fuel chamber and filter chambers inside it is essentially a cylindrical modular device which converts sea water into its individual constituents of hydrogen and oxygen gas.

Water Molecule Power Reactor System is a system that converts hydrogen and oxygen gas and creates a thrust force for use in jet engines turbines that will also produce electricity.

Thrust Force is a force created when a hydrogen gas and oxygen gas is ignited by an open flame.

Wind Turbine Power Generator is a device which converts the power of the wind into a mechanical energy and then into an electric energy.

Gross Thrust Force is a force generated by a device or a jet engine without a load.

Net Thrust Force is a force generated by a device or a jet engine divided by the load of the object being pushed either horizontally or vertically.

Safety Precautions and Warning

The computations presented in this research are experimental in nature with reference to actual experimentations conducted in a safe experimental laboratory with apparatus and devices developed and produced by the researcher.

The data and results derived from this computation were derived through careful considerations about safety, environmental precautions and meticulous planning.

Please consider safety if you are considering repeating the data mentioned on this research. Hydrogen and Oxygen gas are explosive and unstable gases if ignited and excited with an open flame or fire.

Hydrogen and oxygen gas are highly flammable gases which must be given due respect otherwise it will cause damage to lives and properties.

You are advised not to repeat the procedures mentioned in this research for your safety and well-being.

Book References

[1] Young and Freedman. *University Physics with Modern Physics*.10th Edition. Singapore: Pearson Education Asia Pte. Ltd. 2002.

[2] Johnson, Johnson, Hilburn. *Electric Circuit Analysis*.2nd Edition. New Jersey: Prentice Hall, 1992.

[3] King, Caldwell, Williams. *College Chemistry*.7th Edition. New York: Litton Educational Publishing, Inc. 1977.

[4] Webster's Dictionary. *Webster's Universal Dictionary and Thesaurus*. Scotland: Geddes &Grosset 2002.

Internet References

[5] Helmenstine,T. ,"Calculate Root Mean Square Velocity of Gas Particles.",https://www.thoughtco.com/kinetic-theory-of-gas-rms-example-609465?print, 30 January 2020.

About the Authors

Cornelio Jeremy G. Ecle, is an avid research enthusiast and his research interest is in the area of alternative energy source, robotics engineering for medical applications, python and C++ programming, computer interface systems and many other technical research in engineering and allied sciences.

He is a graduate of Bachelor of Science in Electronics Communications Engineering from Cebu Institute of Technology University, Cebu City, Philippines.

He also holds a full pledge Masters in Information Technology from Asian Development Foundation College, Tacloban City, Philippines.

He is also a University Instructor at Eastern Samar State University, Salcedo Campus, Salcedo Eastern Samar, Philippines. He has been teaching in the university since 2012.

For your research related questions please email at jeremyecle2015@yahoo.com.

Cherlowen A. Bolito, is a researcher who conducts technical research in the areas of engineering and allied sciences.

He is a graduate of Bachelor of Science in Civil Engineering from Eastern Samar State University, Main Campus, Borongan City, Philippines.

He also holds a full pledge Master of Arts in Teaching Mathematics from Southwestern University, Cebu City, Philippines.

He is also a University Instructor at Eastern Samar State University, Salcedo Campus, Salcedo Eastern Samar, Philippines.

Jelinda T. Macasusi, is a researcher who conducts technical research in the areas of engineering and allied sciences.

She is a graduate of Bachelor of Science in Computer Engineering from AMA Computer College, Tacloban City, Philippines.

She Holds a Masters of Information Technology from Asian Development Foundation College, Tacloban City, Philippines.

She is also a Faculty Member at Eastern Samar State University, Guiuan Campus, Guiuan Eastern Samar, Philippines.

John Mark N. Gonzales, is a researcher who conducts technical research in the areas of engineering and allied sciences.

He is a graduate of Bachelor of Science in Agricultural Engineering from Eastern Samar State University, Salcedo Campus, Salcedo Eastern Samar, Philippines.

He Holds a Masters in Agricultural Engineering from the University of the Philippines, Los Banos, Philippines.

He is also a University Instructor, Faculty Member at Eastern Samar State University, Salcedo Campus, Salcedo, Eastern Samar, Philippines.

Systenext Alpha 6809 Research Innovations Limited

Copyright © 2021

Since 2012

www.systenext.com

Copyright © 2021 All Rights Reserved.

Since 2008

www.ingramcontent.com/pod-product-compliance
Lightning Source LLC
Chambersburg PA
CBHW070443220526
45466CB00004B/1760